M000286815

i

CEO OF LIVE VIDEO

Discover The Fundamentals Of Dominating Live Video Through The Eyes Of A Producer, With A Step-By-Step Formula To Engage And Convert Your Viewers

DEDICATION:

To the best storyteller known, Father God, thank you for keeping your promises and continuing to write my life story. I am amazed and in awe of what continues to come from this life.

To my family, your support continues to be the wind that propels me. To my husband and children specifically, thank you for always saying YES to every project, speaking engagements, and book without hesitation. What you do behind the scenes does not go unnoticed.

To my purpose pushers, you know who you are. Thank you for the gentle nudge, and sometimes shove towards my destiny.

To every person who has struggled to find their voice in a space already filled with a lot of noise, may this book be the answered prayer for fully showing up with your genius. The world is awaiting your story, your voice, and you no longer have the right to remain silent. Thank you for being part of my WHY story.

CEO of Live Video

Copyright © 2017 by Roshanda E. Pratt

All rights reserved. This book or any portion thereof
may not be reproduced or used in any manner whatsoever
without the express written permission of the publisher
except for the use of brief quotations in a book review.

Printed in the United States of America

First Printing, 2017

ISBN 978-0-692-96726-3

Ordering Information:

Quantity sales, special discounts are available on quantity
purchases by corporations, associations, and others. For
details, contact us below.

www.therosholive.com/contact

FOREWORD:

There are some people in life who just get it. They succeed beyond expectations, take the smallest amount of capital and turn it exponentially and make it look effortless. Those people,take advice from Roshanda Pratt.

Roshanda's years of experience as a TV news producer, along with her ability to create visual stories with words makes her the best in the industry. I've seen firsthand how creative she can be when tasked with the most difficult assignments. She never has to re-invent her formula for success, because she's never seen an equation she can't solve.
This book is the reader's first step into creating, writing and broadcasting their "get it" story!

-Darci Strickland

Darci Strickland is an award winning journalist with more than 20 years' experience. She's been recognized nationally for her investigative reporting. Darci has a unique way of crafting stories that not only make you think but also feel!

Table of Contents:

My CEO Story:

It's time to show the world who's the boss!

World domination can be yours! Okay, maybe not world domination, but you can at least conquer your world through live video. What does it mean to be the boss or the CEO of live video? Let's begin with the business definition of CEO. Investopedia, the world's leading source for content online defines CEO as, "The highest ranking executive in a company whose main responsibilities include developing and implementing high-level strategies, making major decisions, and managing overall operations and resources."

Live video is a strategy that continues to grow in popularity across several digital platforms. According to Marketo, users are ten times more likely to comment on video than on other forms of media. I think the real reason is that live video is in real time and has real interaction. People want a relationship with brands. Don't think so? Here's one of many stories I will share. I had a bad customer service experience with one popular brand; I won't share the name to protect the guilty. After months of trying to rectify the situation, I went to Twitter. The company responded with a direct message, and the problem was solved. On the flip side, I had a wonderful stay at a hotel and applauded them for great customer service on Twitter. Their social media manager responded with a thank you and added points to my customer card. It was a totally unexpected and welcome surprise. In these examples, the companies demonstrated active social listening. Why? Because both understand people want a chance to be seen and heard. People want what we were all created for: connection.

How does this relate to live video? To be the boss of your marketing efforts, your storytelling, and your brand, you must become a CEO. In this instance, CEO is an acronym for Connect, Engage and Offer. While the main point of the live video is to educate, engage, and encourage, there has to be a connection and intentional engagement, which leads to an offer for your audience. As a CEO of my live video, I've learned how to use this high-level strategy to gain more clients, increase my influence, and monetize my brand. Live video has given me the leverage to go across the country and internationally without leaving my home physically. Live video has allowed me to increase my bottom line, work with clients who also want to OWN and dominate in the online space. It has also given me an opportunity to build a media platform. This did not happen solely based on my television news career. Although it helped, I had to learn how to show up in a way my audience loves. And trust me friends, your audience loves live video. With 20 years' experience working in the television media, I know the power of media. Words and pictures make the perfect marriage. Storytelling pulls at the heartstrings and prompts action. But live video is much different than what we do on television. As broadcasters, we talk at the viewers and give them information. In live video, we speak with you, and not at you. There is a resemblance and yet, a great difference.

As the CEO of your brand and the rising CEO of live video, my desire is for you to put video communications at heart, and focus on your business. It's really a low hanging fruit, and the point of entry is minimal at best. You can no longer afford to ignore this platform. Whether you are a novice or pro needing a refresher, live video is here to stay. A Native American Proverb says this," Those who tell the stories rule the world." Those who are telling their stories, whether true or false, those taking the plunge to hit the go-live button are those who are ruling. They are the real bosses!

Don't you think it's about time your status goes up?

Are you tired of hearing crickets during your live video?

Do you feel disconnected instead of connected to your live audience?

Do you desire for your live video to have a greater impact?

What if I told you all of the above is possible.

Technology is constantly evolving, and changing from the printing press to the telegraph to the telephone, and then the upgraded smartphone technology gives us a new way to communicate. Live video affords everyone, and I do mean everyone the chance to become a CEO. Long gone are the days of waiting for traditional media to give you a chance. Nope, you can create your own media empire.

Your story deserves to be told. You need to be heard. It's time to connect, engage and offer BIG like the CEO you are born to be.

This is your time to make some Boss moves when it comes to your visibility and creditability.

This is the CEO live video life. Welcome.

Meet The Rosho:

Why should you listen to me? Well, let's begin with how I got the name The Rosho? It came from my more than 20 years' experience working behind the scenes as a television news producer. My fellow producers would joke as I would go into the control room. It's The Rosho. Well, it stuck and followed me at the numerous stations I worked.

My love for media started in the fourth grade. With only one television in the house, my West Indian father would make me watch the evening news. While my friends were watching the Smurfs or Scooby Doo, I was well versed on current events delivered by Peter Jennings, Tom Brokaw, and Walter Cronkite. My father told me as an American born citizen; it is my duty to know the stories of my community. Storytelling is in my blood. Connecting and engaging with people is my mojo.

As you read this book, I am your new business partner. My clients know me as the storyteller strategist and heartfelt producer. From a television producer covering breaking news stories to crisis communication, marketing and on both sides of the camera, my super power is helping influencers just like you discover their brand stories to connect, and impact the clients they are called to serve. I've had the pleasure of working with The W.K. Kellogg Foundation which entrusted me to drive one of their national campaigns. I'm also a two-time #1 Best Selling author, Huffington Post contributor, and host of my own self-produced Facebook Live Show, "The Rosho Live" I appear regularly on my local CBS morning news show and have been featured nationally on CBS' The Talk, numerous podcasts, and telesummits.

From my personal experience of building a brand to the work I've done with dozens of entrepreneurs, I've learned the secret sauce to live video, and I am excited to share it with you. If you are ready, turn the next page to get started.

Chapter 1: Making a love connection

In the 1980's, I loved watching this show called 'The Love Connection.' The dating game show hosted by Chuck Woolery, attempted to connect singles with a compatible partner. The show was really ahead of its time using videotape profiles for a guest to pick one of the three contestants to go out on a date. The Love Connection understood something we are now discovering years later. Video is a powerful tool to make a connection.

More than a decade working behind the scenes and in front of the camera, television news taught me the power of the lens. There is something magnetic about video. In this case, when I am speaking about video, I am also speaking about live stream. In 2016, there was a resurgence of video. In the television industry, we have long discovered the impact of words and pictures. I often say the two are the perfect marriage. With the creation of such apps like Meerkat, Periscope, Blab, and Facebook, live stream puts everyone on the same playing field. No longer do you necessarily need traditional media for your local news station to become a media personality or powerhouse.

Here are a few stats to consider:

1/3 of all online activity is spent watching video

92% of mobile video consumers share videos with others

According to Hubspot, the average online user is exposed to an average 32.3 videos in a month

Stories are 22 times more memorable than facts alone

Our brains process visuals 60,000 times faster than text.

As you can see, people love video and according to the industry, video will continue to dominate as the main way to connect with your ideal client/audience. Why are people fascinated with video? Here are a few reasons: Video gets your attention—it's the perfect marriage of words and pictures. Video gives you the ability to explain complex concepts clearly and quickly. I am amazed as a journalist, how we can take a complex issue like taxes, and break it down in a less than 2-minute video. Video has the power to trigger emotion through the art of storytelling. Need a few more reasons? Video allows you to take people behind the scenes or behind the curtain for a real one-on-one connection. Finally, you instantly become a speaker and have exposure to an international audience without even leaving your home or office. Video is honestly a win-win

So, tell me again why you are not using live video? Are you struggling with fear or perfection? I'm here to tell you, your audience does not want perfection, but they want YOU instead! No, seriously, perfection is overrated. The urge to want to have a perfect video truly is counterproductive to what your audience desires. My friend who's a brand manager at one time for a high end sports club in the DMV area shared a story with me about video. For a period of time, trainers at this club would post quick training videos showing clients exercises they could do outside the gym or share gym techniques. The club realized this was a booming success, and wanted to capitalize on the growing trend naturally. The company hired a team to come in to create some branded videos. With lights, cameras, and action, they spent thousands creating training videos, and the result was crickets. Their audience did not respond. The videos were perfect, great lightning, editing, and all the sparkle and shine, but their audience did not respond as they did to the other videos shot by the trainers on their smartphone.

So, what was the disconnect? Although, the videos are produced well, the fitness company realized they did not work with their audience because they were "too perfect." Now does this mean you should not have pre-produced, edited video? NO! But knowing how you will use them matter. For example, the welcome video on my website, therosholive.com was pre-produced because of the purpose. But all of my videos to connect with my audience are in real time, live. Think of it like this- when you connect with people, you do it in real time. It's not a recorded conversation, and it is not always perfect, but it's human and real.

How Do You Make The Connection?

Making the connection begins with understanding you have three things: You have something to say, creating valuable content, and telling a story. Let's begin with the understanding that you have something to say. I often hear clients tell me, Roshanda, I would love to do live video, but I don't have anything to say. My friend that is not true. You are not barren of information. But it begins with understanding your voice matters, and you should own it. I remember in elementary school coming home with my progress report card. My West Indian father looked at my teachers comments. Overwhelmingly my teachers said I was a good student, but just talked a lot (No, surprise there, right?). My father turned to me and asked, "What do you have so much to talk about" Without hesitation, I told my father I believe I have something to say. How about you? Do you believe you have something to say? You no longer have the right to remain silent. Today, I am giving you the permission to speak up because what you know is worth something. You are valuable!

Think about YouTube; from cat videos to how-to videos, YouTube's fame is its ability to give everyone the opportunity to share what they know. If you look on YouTube, there is an abundance of videos teaching you how to do a Smokey eye, although, I still have not perfected it. You can share your education, your expertise, and everything in your 'noggin' with your audience. We live in an information society. You, my friend, are the solution to someone's problem.

Let me say that one more time if you are feeling depressed today. If you are feeling down in the dumps and having a low self-esteem moment, let me inform you today: You, my friend, are the solution to someone's problem. You are the help to their challenge. You should say that at least once a day.

Today, I am someone's answer, and you are. And that's the joy that you get when you show up on live video. You get the opportunity to solve someone's problem. Your main goal when you show up to be the answer to their challenge, which means that you have to come here ready to provide value and inform people. So when you show up on live stream, you have to show up ready to provide valuable information. No fluff. People can get that from Google or your competitor. Your competitor is giving them fluff, and not you, my friend. You're going to give them tangible and great information that's going to change and impact their lives.

OWN YOUR VOICE!

Connecting the content:

Where do you begin connecting the content? A CEO of live video begins with their audience in mind. After all, the real star of your broadcast is the people who are watching you. If you don't know where to start; begin with asking your audience what they want. Do you market research by posing questions? Open up a dialogue with your audience and let them know you are listening. Consider this; when you go to a restaurant, and on the back of your receipt or at the bottom you are invited to do a survey. Why? The restaurant you just dined at desires your feedback. Allow your audience to give you feedback after they have feasted with you.

Ask yourself what do people constantly ask me about either face-to-face or by email inbox. Take a look at things happening every day, and create content around it to educate, inspire, and prompt action. I've done that a lot of times on Periscope, where there are things that are happening in the media. I use those examples to teach my audience how to engage with the media correctly. What never to do or how you can take this story and insert yourself into the conversation. So pay attention to the world around you, and use that as a tool to educate people as well.

So one of the ways that you can provide education and information to your audience is, are you ready for this? Share something that you learned from someone else. I know. Shocker! Right? Because some people are like, is that copying? Is that, taking somebody else's information? No, not if you're giving them proper credit. I'm going to give you a perfect example of that. Not too long ago, I did a Facebook Live in my FREE Facebook community, The Rosho Live Media Mastermind group, discussing the three videos that you can create today. One of the members of my Facebook community took the information I shared, and instantly went on Periscope and shared the same information, and during her broadcast, she gave me credit several times. It was a great live video. Here's what happens. Not only is she providing information and education to her audience, but she's also showing her audience that she's open to being teachable. That she is always learning her craft and studying her craft, and she's bringing them new information that is hot off the press. The other thing that she did was that she opened up her community to my community. Now, her community is like, WOW! That was really great information. I need to follow The Rosho Live. It's a win-win.

Now I shared what you can do, let's talk about what you should not do. Do not share what your clients pay for to your audience for FREE. That means if you charge a client for a VIP day, and you share certain information with them, you shouldn't get on live stream and then share that information for free because it's not right.

Here's why: You're charging them for this information, but you're willing to show up online and give it to the world for free. I see this happening a lot on social media. The main reason I think it happens because every human has this need deep inside to be liked and accepted. There is something which sparks on the inside, some scientists even have compared it to the drug high when we see people liking, engaging and even sharing our post. We want to make sure that we're giving people valuable information. We want to make sure that we're showing up for people. So then we put all of this information out there, and we give people everything in the kitchen sink. And then it's like why should I work with you? Why should I bother signing up for your newsletter? Why should I bother reading your book? You just gave me all the information in the book.

Now, does that mean you play peek-a-boo with your audience? No, that's one thing I learned as a television news producer. You know, there's a thing in the television industry called a tease. The objective is to entice the audience about what's coming up, so they stick around and not switch channels. You don't want to play peek-a-boo information with your audience because they're going to get mad, and you don't want to promise them something, and then you don't deliver. So you want to be able to give them tangible information that is just enough to peak their interest but also provide them with valuable information that they keep coming back. I've been there where I want to give value and make sure people come back, so I give, give and give. But I also learn you are shooting yourself in the foot because when you give your people all of this information, several things happen. If you are engaging new viewers too much information, you'll cause them to do nothing. It's almost like you just brain dumped all this information on them that should be in an eBook. And now they are just like, Well, now what do I do with this? And they are overwhelmed and confused, and remember, overwhelmed people do nothing. You want them to do something. You want them to come back to your live stream. You want them to sign up for your coaching program. You want them to sign up for your e-mail list. You want them as a client, especially if they are your ideal client.

The other thing which happens when you put too much information out there is that you train your audience to always expect a freebie from you. You're scratching the surface. It's like toast that's burnt. You are just scraping it off, but they are not getting any real substance. Don't waste people's time. Give them substance, and there's a way you can give people substance and still sell while you are giving them substance. Give people the meat and the potatoes, and stop giving them a salad. Salads are good, but if you are hungry for a steak. How many of you know a salad is not going to work? So don't give people lettuce and then tell them this is a steak because people are going to be mad. Don't upset your audience with broken promises and fluff. If you do not deliver, they will go elsewhere to get it. But there is something you have your competition can't beat you at, and that special something is your story. Marry your content to a compelling and authentic story to become a real boss!

CEO Work:

List out 5-10 Broadcast topic ideas

1._____

2._____

3._____

4._____

5._____

6._____

7._____

8._____

9._____

10._____

Connecting with a story:

Facts tell, but stories sell. My favorite Native American Proverb states, "Those who tell the stories rule the world." Don't you think that is true? Seriously, take a look at the world around you. Whether the story is true or false, we are see how stories are overriding facts every single time. Each day we are telling ourselves a story or being sold one. Your audience desires a story too. Since the beginning of time, we humans have been telling stories because we connect to them. The hieroglyphics of the Egyptians, The Greek stories handed down through the centuries, biblical stories, and yes, even the stories in your family. Stories record history, give us wisdom, and set rules for moral behavior. Stories are what make you human.

Your story gives people insight about who you are. In this relationship marketing era as my dear friend Dr. Fred Jones says, "Transparency is the new currency." Tell more instead of selling more, and you'll sell every time. The art of storytelling build **KNOW, LIKE** and **TRUST**.

We have to ask ourselves a few questions? Why do I want to use live video? Who am I speaking to, and what do I plan to say?

Listen, we have enough Facebook fakers and Periscope phonies. We want our live video to have purpose and impact. If you are doing this to become popular, then you are missing the point. No, instead think of it like this, I show up on live video (noticed I said show up) because I want to be relevant first. I believe true impact which leads to "popularity" or being in high demand comes from understanding relevancy over a few likes and hearts.

Now you're probably asking, "Well, Roshanda how do we get there?" I am so glad you asked. It starts with understanding your vision and mission. I think for so long we have left those terms just for the business or no-profit world without realizing you as a personal brand, personality or business owner needs a vision or mission statement.

Your message or story must be told like a mission. While keeping the vision of where you want to take your audience in the future is before you.

Mission versus Vision:

A mission statement defines the present state or purpose of an organization; it MUST answer three questions.

WHAT it does; or **WHY** it does it
WHO it does it for; and
HOW it does what it does.

Notice **HOW** is last? Too often in the storytelling process, people want to start with the **HOW,** but what people want to know first is the **WHY**.

Here is an example of a mission statement:

Target: "Our mission is to make Target the preferred shopping destination for our guests by delivering outstanding value, continuous innovation and an exceptional guest experience by consistently fulfilling our Expect More. Pay Less ® brand promise."

A vision statement defines and gives a mental picture of what will be achieved over time. It's the 'north star,' the compass, the direction for employees and clients as to where you are going.

Here is an example of a vision statement:

Alzheimer's Association: "Our Vision is a world without Alzheimer's disease."

Avon: "To be the company that best understands and satisfies the product, service and self-fulfillment needs of women - globally."

Microsoft: "Empower people through great software anytime, anyplace, and on any device."

A few questions you can ask yourself: How will we make a difference every single day? How will we make the world a better place through what we do? These questions do not center about popularity or making more money. However, I believe when you put these things at the forefront, you will be able to monetize your story and amplify your voice.

When I created my first digital product in 2015, I had no clue what I was doing. I followed a popular live streamer who encouraged me to create a product simply sharing what I already know. I could not believe friends; it was that easy. Well, I created a product about how to pitch the media and released it on Black Friday. Friday came and went, and Saturday evening I had no sales. Honestly, I felt dejected. I spent time creating this digital audio and PDF download, and it would appear no one purchased it. I decided to keep it up. On Sunday afternoon, I was watching popular live streamer and branding coach Shade Y. Adu on Periscope, speaking to well over 100 people. She noticed I came in and stopped the broadcast to address me. She said she purchased my product. I was in total disbelief. She also went on to say it was well done and was so valuable, so I should charge more. She spoke about my background as a television news producer and my mission to help people find their voice, clarify their message, especially to the media. Instantly, people started asking for the link to my product and started connecting with me on Periscope WOW! I was blown away. Just imagine if I took the product down? It has become one of my company's best sellers. My main objective for creating that product was a challenge to help people failing, and also to make money. To this day, that broadcast opened the door to great relationships and faithful viewers of my broadcast because my motto is to lead with value always, and stay true to the vision of raising messengers!

CEO Work:

What stories can you use to pair with the above content? (refer to broadcast topics)

1._____

2._____

3._____

4._____

5._____

6._____

7._____

8._____

9._____

10._____

How do you become a better storyteller or seller?

Remember what I said, every day you are telling yourself a story or being sold a story. Stories are what connect us. Here are three ways you can become a better storyteller or story seller.

1. **It starts with telling a narrative**—a story about how the product and service fits into the fabric of your client. It's about being the solution to their pain point or their life in general. Storytelling gives

your client insight into why they should pick over someone else.

2. **Your clients have a lot of "noise" to contend with**. What will separate you from everyone else? Is the problem and the emotion. Far too often we avoid this. Now when we talk about emotion this should be to benefit the listener, and not make them cringe or uncomfortable. We don't vomit our emotions on people, but we connect them to it. That means you have to be comfortable with going there. What makes your clients fearful, feel loved, hopeful, and what do they hate? These are all emotions you have to understand where it fits in the story, and tap in a real and simple way.

3. **A critical part of story-telling is transparency**. Why do we wait until something bad happens before we take people behind the scenes? People want to go behind the curtain. People desire to know where the idea from a product came from, how it's made, why you're doing it, and the people who have used it and so on.

Bottom line: BE REAL!

It tells a narrative of how the product fits into the fabric of your consumer's everyday lives – not just the pretty pictures, but also, the honest way people interact with your product. It's weaving your brand into the relevant passion points of your ideal client's life.

Things to Keep in Mind When Story-Selling

A great story seller begins with a hook that entices curiosity. Open with a statement that will make the user want to hear more.

Keep your story minimalist in its form. The more longwinded your story is, the more chances your listeners' attention will wander.

Contrast a clear "before" and "after" situation in your story to show the user the benefits that your product brings with it.

Make sure you have a happy ending. Biology supports happy endings in stories by releasing the neurochemical dopamine into our limbic systems that leave us feeling good about ourselves and the brand associated with the story.

A great story does not just live inside best-selling novels or Hollywood blockbusters. Every brand carries a story hidden inside of it. All you need to do is draw it out and retell it to your audience to let them make your brand their own.

CEO Work:

Take inventory of your story. What it offers, include visuals to support, analogy, personal story and a client's story.

_____

_____

_____

_____

_____

_____

_____

_____

_____

_____

_____

_____

_____

_____

_____

_____

What are the stories you should be telling?

The story you should be telling is the one which connects you with your audience.

How well do you know your target audience?
Do you know what pains them?
Are you willing to be vulnerable?

Let's discuss being vulnerable. What it is and what it is not? If transparency is the new currency that means my transparency has to come from an authentic and healed place. We have too many people spilling the beans, but you can tell they have not gone through their own space of healing. What do I mean? They realize they have a story; they want to share it but are willing to do it at the expense of still grappling with their pain. I've seen plenty of people cry on a live stream. However, there is something to be said about the person who weeps as they still relieve the pain of their past which is still difficult to talk about in front of people. You have to be well. Now, am I saying you should not show emotion? No, emotion makes you human. People connect to emotions. Just think about your favorite movie. There is a reason why it's your favorite because there is a connection to the story, maybe even an emotional one.

As a once television news producer, I was taught early on not to take your viewer on an emotional rollercoaster. For example, to go from bad news like shooting to happy news, a family reunited, and sad news of a fireman dying. I'm telling you, don't take your viewers or listeners on an emotional rollercoaster. Be intentional about what you will share, how far you will go into your personal story, and how.

You should know your target audience so well you can write a page in their journal. Understand their pain, their fear, hopes and hate. This helps create a better program to solve it, and it helps with marketing language and media pitches. It helps with creating a connection that connects and converts on live video.

Think about your last purchase. I guarantee you told yourself a story. For me, it was YES, to another leopard print dress. I knew I didn't need it, but I wanted it, and told myself I could wear this to one of my speaking event coming up, I told myself a story whether it's true or not. Your target audience has a pain point you are called to solve. But you must understand their pain, the language of their pain and how to connect to it. For example, my audience is pained by not being heard. They know they have a story; a message but not sure of the words. They face fear, doubt, and want to show up authentically and compellingly. I know my audience pain because I've once faced these same struggles. Imagine working in the field of communications, writing the words anchors, read on air, only to be shaking in your boots about your voice? It was a torturous existence. Once I discovered my voice, I was determined to set others free. More likely that is a bit of the story of the women and a few brave dudes, I work with. Connecting your story, your client's story to your product, service or brand allows your client to see themselves in what you are offering.

So after you discovered the story, how do you create the content? We have all heard content is KING, and it is a way to connect with your audience. My first years working in the television industry were spent working the overnight hours on a 2.5 hour morning show Monday through Friday. I would come in at 10 pm and work to about 10 am. The hours were grueling, but it taught me a lot about creating content. Just think 2.5 hours is a long time to fill with news content especially on a slow news day. We had to be creative, resourceful, and learn how to repurpose content already used. I know this too well, and use it in my business to teach clients how to do the same. My friend, you have more content than you can imagine. I am often do ask about what to say while on the live stream. My answer is usually the following:

- **FAQ: Frequently asked questions:** What are people always asking you? You can video or live stream the answer.
- **Behind the scenes:** People love this. Take people behind the scenes of how you are planning an event or the event itself. Show them how you are working on something
- **Master training:** We are in an infopreneur space. Show and tell. Teach people something that provides benefit.
- **Interview:** As technology continues to grow there are several ways you can interview an industry influencer or client.

The list goes on and on about what you can provide as a valuable content to your audience. Remember your formal education, things you learned from a mentor, training, book knowledge and street savvy can all be things you glean from to create content your audience will devour. When I am speaking to Public Relations Professionals about pitching the media, I tell them to make the pitch yummy like a Krispy Kreme Doughnut. If you never had a Krispy Kreme Doughnut, you are missing out. The hot sign is one letting you know the doughnuts are fresh, and when you take a bite, it melts in your mouth. Your content should be so yummy to make audience ask for more. It is so yummy they can't help but comment, like, heart and even share. Your content prompts action. The action is what you want because it means you are engaging your audience. In the next few pages, we are going to spend some time talking about the power of engagement. I will share what has worked for me and many others you watch utilizing the live stream to build their brand platform.

Chapter 2: The Power of Engagement

I remember the first time I "tried" live video to an audience I had not built know, like, and trust factor with. It was a scary feeling. There I was just talking to myself, looking at the numbers and seeing none to very few people watching me. Honestly, I was slightly intimidated. However, deep inside, I knew this platform was a game changer. Those first months served as a life lesson on what to do, and what not to do. Let's begin with the word **ENGAGEMENT**. Simply put, I define it when it comes to live video as the act of getting to know, building a relationship. How well do you know your audience?

If you are on live video and not trying to get to know your audience or allow them to engage with you, then what is the point? While live video reminds me some aspects of television news where they differ is the engagement piece. The anchor reading the 5 o'clock news is doing just that, reading. You don't get a chance to ask questions in real time. What you see on television is what you get. With live stream, your audience can ask questions, and to talk back to you. One the broadcaster is talking at you, the other allows the broadcaster to engage with you. You must understand engagement leads to profits. I've seen many broadcasters who give out information, and then leave live stream or end the video abruptly as though, they are afraid to talk to the people, or even to show people who they are. Long gone are the days of marketing products, we are in a relationship building era. Your network determines your networth. How you engage, and intentionally build relationship determines how much you make.

Digital Marketing experts say most Americans are exposed to 10,000 advertisements or messages daily. If you are like me that number can be overwhelming, the first thought is, how do I engage when there are so many voices? What makes the difference is, understanding who you are called to engage with. I'm not called to everyone and I'm okay with that. There was a time I was not. If I am honest with myself, it's because I did not understand who I was called to connect and engage with. Here's a secret: when creating videos or on live stream I think of my ideal client. She has a name, and I imagine myself talking to her and not the dozens who are watching live. It helps me to engage better, and no wonder, those who watch me on live stream say they feel like I'm speaking to them directly. If I could, I would relate engagement to a Krispy Kreme donut. I know, I relate a lot of things to donuts. But the glaze on top is your engagement. The bottom line engagement is more about building a valuable relationship. As the sign says in my home office, "I will solve your problem, and you will pay me.

Back to my Krispy Kreme example, when the hot sign is on, you better believe those donuts are solving a need. Every time that hot sign goes on Krispy Kreme is engaging its audience, saying the donuts are hot, and we are ready to serve you. Your video or live stream is much of the same. Each time you hit the record button or Go Live button, you are telling your audience I'm ready, and I'm going to serve you some hot content. This is why I tell PR Pros, and I'm telling you to make your media irresistible and yummy like a Krispy Kreme Doughnut. But to make it tasty, you've got to have an intentional plan.

Plan your Engagement?

Michael Hyatt says, "If you don't have a plan for your life, someone else does." If you don't have a plan of engagement, your competition does, and that is where your audience is going. Confused people do nothing! This is why you can't just have good intentions because good intentions are good for nothing. To be a CEO of live stream, you need to connect, and the second tenant you need to engage requires a plan.

I had the opportunity to work with the W.K. Kellogg Foundation for their National Day of Racial Healing campaign. It was an invigorating experience. My role was to help communicate the foundation's message of creating a day where people had real conversations about race. I realized that to get people to share their story they had to "see" someone sharing their own. Humans connect to stories they can especially identify with. To help the foundation connect their message, I created a video campaign where I shared my story of being an American born daughter of Caribbean parents and the stigma I endured in my community. What happened next surprised even me. My story resonated with so many people with well over 2k views. You can check out my story on my YouTube page.

My story of racial healing jump-started a discussion online, and in the Caribbean community which until this point felt left out. As a storyteller strategist, I never considered this community as silent, and I never considered myself the voice. But this is what happens when you intentional tell the story you uncover another segment of the population you never considered prior. For the W.K. Kellogg foundation, the engagement opened up opportunities for media interviews. I had the chance to do one of those interviews with a New York radio station talking about the foundation's efforts, and share my views on race relations in America.

Engagement is the prerequisite to action. Think about a car salesperson. When you arrive at the dealership immediately, they start engaging with you. They start with asking questions. The objective is to learn more about you. Why? So they can now build, know, like, and trust with you to lead to a sale. It is all planned. My husband is licensed to sell life insurance. He studied to take an exam to give people the right information and to be in regulation with our state laws. As part of his training, he also attended what they call sales school. During these classes, he learned how to engage with people and much of what he learned his asking questions Not only are agents given a script of sorts, but it is planned how they engage with their client. There is a certain level of intention. The purpose of engaging is to turn those viewers watching your broadcast into raving fans part of your Tribe, and those who ultimate buy from you.

Rules of Engagement:

I learned the rules of engagement while working in a television newsroom. Every day during the morning planning meeting while we would discuss ideas of what stories to cover we also discussed how to engage with our audience. With the invention of social media, this question became more paramount. Things like creating Twitter polls, Facebook questions, encouraging posts where we could share viewer's pictures or comments have become the norm. Engaging with our audience creates brand loyalty, allows our audience to help shape content, and connect on a deeper level.

There are a few things to consider if you go live. What's the point of going live if no one knows about it? Before you start streaming, make a plan; a plan to promote. Post photos or video alerting your audience you are going live and when. In television news, it is called a TEASE. This is when you give an enticing tidbit about your broadcast. When promoting, go beyond the current platform and do multi-platform outreach, inviting email subscribers, blog readers, and other social media followers.

It's only a test; take time to prepare by testing your audio, lightning, etc. Nothing turns an audience off or causing them to switch the channel is dealing with technical issues that could have been prevented. Test your wifi connection, so your video comes across crisp and clean. And remember, if you are using a mobile device, set your phone to "do not disturb" or "airplane" mode as you are not interrupted during your broadcast. Also, part of the pre-show preparation is having your content in place. What do I mean? When I am preparing my notes, I make sure to have a very catchy title, three talking points and my impact story. I'll get into this more a little later. However, in my television news days, I was taught viewers could digest what we called then the rule of threes. So, I still keep my points to three to make sure I am providing value, viewers get the most out of their time with me, and to hold myself accountable not to give away the kitchen sink. It's easy to get caught up in the likes and hearts, and just start rattling off information because the adrenaline rush of being like feels nice. Do not fall into that trap.

Turn up the energy. From the start of the broadcast smile and give your audience energy. Make it hard for them to turn away. There is something to be said for a warm and welcoming smile. If you deliver personality and energy, your audience will respond back with the same energy and invite others to the party. Remember you are taking your audience behind the scenes of the brand, this is a rare opportunity for them to engage with you one-on-one, use it to your advantage.

Do a roll call. One of my biggest pet peeves is people who get on live video and say the following: "We are going to wait for more people to show up before I start." Here is what you are communicating to those few people who decided to join your broadcast. Sorry folks, you are not that important until I get more eyeballs. These are the same people, who are investing their time to join your live video; please do not punish them. Instead, take the time to introduce who you are, how you serve the world and say the name of a few people whom you see in the comments section. People love hearing their name called by their favorite broadcaster, make your audience feel important. Make the show about your audience because they are the real stars.

Do reset. Remember experts suggest lived video going at least 15 minutes. There is plenty of time to reintroduce yourself as you see your live audience growing. Viewers may be joining throughout the live stream, so refer back to the topic of your broadcast. This is also a great time to remind viewers to share your broadcast. Sharing is caring. My mentor as a young producer taught me and said, tell your audience what to do, or they will do nothing. Encourage them to share the post with their audience, to follow your page and comment. Think about holding a specific time your audience can always count on seeing you live. Much like watching your favorite program, you know when it comes on each week, clue your audience in on when you are coming on live.

It's never really over: After the broadcast, this is where the real work of engagement begins. Go to the comments section and make sure to respond to questions and comments. Thank people for watching, and again ask them to follow up and share your broadcast.

Your aim is to become an influencer to your audience. You want to become their go-to resource or expert. It's what I like to call **Instant Celebrity Engagement**. It is the thought of always thinking of ways to create engagement which garners a response. Have the consistent habit of showing up to your audience and watch; they will show up for you.

Here's a secret, are you ready for it? Television Networks understand the power of engaging their audience beyond the TV screen. In fact, they want to win viewers on the big screen, and the small screen of your mobile device.

Essentially, television stations want to engage you on both media platforms, and it is understood most of their audience walks up in the morning and checks their phone. However, there is still a segment of the population who are regular television watchers. Why am I sharing this with you? I want you to take a page out of the television networks handbook, and realize you too are a media warehouse. Consider this: You audience watches you because they trust you. This is a great responsibility. This means we do not have the luxury to be haphazard with our platforms. We have a social responsibility to post truth, to be authentic, transparent, and hold others accountable to the same standard. We engage on purpose for a purpose. Your main goal is NOT to be popular, but to be relevant. Be relevant, show up consistently, engage intentionally and give value. When you do these things, you'll always be in demand.

BONUS: Live Video Best Practices

While we are discussing engagement, I wanted to share a few of my best practices for live video which make me a BOSS of the screen.

Attention Grabbing Title: Engage your audience immediately from the title. Get them interested, and give them a reason to listen.

Here's a quick formula: What you will share + outcome = title.

For example: 3 ways to get viewers never to leave your broadcast. See how I gave a promise of 3 tips on what will be shared and then the outcome? Who wouldn't want this? One more thing I should mention, your title should be inviting, catchy and not a lie. Don't promise something and fail to deliver.... that is one way to lose creditability and an audience. It's never worth it.

Powerful Introduction: Introduce yourself from the start. Tell people who you are, and how you show up or impact the world. This should be in an elevator pitch form, short and concise but powerful. It should include your life's work and who you work with. Do not make it stuffy! That's a No-No. Show some personality, if I may use this thought: Blind your audience with your shine, they can't turn away.

Know Your Destination: Decide before you hit the Go Live button where you want to end. Think end before the beginning. Where do you want your audience to end or leave with?

Tell the Story: Build audience engagement by making your subject immediate and personal. Facts tell, but stories sell. Mention your experience as it is appropriate to the subject matter. You can even use client's experience when necessary.

Close like a Boss: Thank all who came, and let them know when you will be back on (build anticipation). Give a strong close, and remind them they should follow you to get instant updates on your next broadcast. Give them an offer (we will talk more about this in the next chapter).

Repurpose Content: Nothing is wasted, and I do mean nothing. Repurpose your video. Download it and turn it into other pieces of content. Post a snippet on your other social media profiles, take a chunky piece and add to a blog post, take a piece of that really awesome moment during the live stream, you know the part where you are dropping some serious nuggets and post on YouTube. Think beyond just the live video, and how you can get more leverage out of your hard work. You may even want to turn into a digital product (wink, wink).

Chapter 3: Make Me an Offer

Live video can remind me of the game show "Let's Make A Deal." Much like the show you are competing in a way for cash. In this chapter, we are going to discuss the taboo subject of making an offer on live video. Now back to my example. On the game show "Let's Make a Deal," contestants win cash or prizes by choosing a certain number 1, 2 or 3. But there is a catch; contestants are tempted with prizes in small boxes or a wad of cash in front of them. The contestant could win something bigger or get a whammy of a prize like a pet pig. You are probably thinking what does this have to do with live video. You MUST make an offer, or at least, tell people what to do at the end of your broadcast. Make a deal with your audience.

Here is the sobering truth: If you don't give your audience an offer someone else will, and they may end up with a bad deal. In this era where everyone is a coach or promises something they can't deliver, it's playing Russian Roulette with your business, purpose or service. Real talk here, I can't begin to tell you how many have told me they would like to work with me, but have had bad experiences with others. They are scared to make another investment and rightfully so. As I tell my clients, when you don't show up on video, you subject those who you are called to serve to the wolves. I don't want that for you. But I also know from experience, when you don't offer your audience a chance to go deeper with you they will go elsewhere, or they will stick around for the "**FREE**" information and never commit.

Your broadcast should end with a call to action or CTA. A dictionary definition states a **call to action** is *an exhortation or stimulus to do something to achieve an aim or deal with a problem*. As it regards to marketing, a CTA is a piece of content intended to induce a viewer, reader or listener to perform a specific act, typically taking on action steps or some instruction. In television news, I learned as a producer you have to give your audience instructions on what you want them to do next. If you watch your local evening news, there is always a push to read more on their website, follow them on social media or call them with a story idea. These are a few examples of a call to action. You want people to take action and not just wait. People, who sign up for your newsletter chose to follow you, head to your website to do that because they are so *moved to do so not as a* suggestion, but as a high recommendation. You are the solution to your audience's problem. Don't be afraid to tell them they need you, and how you can help. Like a fisherman on the dock, the call to action is the hook, your live video is the bait, and your ideal client is the target. Trust me, we are all trying to hook the right client, but that does not happen until you have the right hook. No more live video without a call to action. No more live video where you end it abruptly or awkwardly try to tell people what to do. People need next steps. Now understand your call to action does not always have to be monetary. I think the 80/20 rule used in social media can be applied in live video offers as well. The rule is 20% of your content should be used to promote your brand, and 80% is dedicated to content which interests your audience. In your call to action, 80% should focus on audience improvement and

20% pushing monetary investment. It's about building community; The KNOW, LIKE and TRUST before asking to get into their wallet.

Before we move forward, we must confront the blaring pink elephant in the room. You know the one who says this is uncomfortable to talk about money. It feels sleazy and inappropriate. Here's the deal, are you in business to make money? Do you want to impact more people? Do you want to grow a brand? Truthfully, I hated calls to actions. It felt uncomfortable, and in my heart, I want to give people information to be better. But I wasn't making any money. I was frustrated because I felt used by the people watching who demanded more but were not willing to commit. I felt dejected. I took a break from live video to figure some things out. I discovered I was the problem. How was I letting people know how we could work together? Honestly, I wasn't. Like many other broadcasters during that time, I got caught up in giving away the kitchen sink. There is something almost, dare I say it, magically about seeing your live audience engaging with comments, Emoji, and likes. It can be compared to a high. But after I came down from the high, my bank account and wallet was still dry. I had to come up with a strategy and one quickly where I did not fall into the trap of giving it all away for the sake of being liked. Do you struggle with that? The best cure is to start with the end in sight. Ask yourself where do I want to end? How do I need my audience to commit? To do this, you must finish your video strong. Here are a few ways to do it.

Different kinds of calls to action or CTA:

Your CTA or Call to Action does not have to be the generic call me or email for more information. It may work, however, think creatively.

Formulate a Question: Invite viewers to "join the conversation" on your brand's platforms or using a dedicated hash tag. For example, during my Live Video, I encourage watchers to post their questions using #ASKRO. Then in another live video, I use that hash tag in the title, mention the person who asked the question (of course, asking their permission first).

Join the Group: Invite viewers to become part of your tribe. If you have a community or membership group, tell them the benefits of joining and show them how to join.

Sign up for a webinar: If you give webinars on a regular basis to showcase your expertise or products and services, ask viewers to sign up for the next webinar. Remember to mention the link, and post in the comments section.

Direct to another video: After your video, invite them to watch another on your YouTube channel or a different version of the same topic.

Power of the Vote: Ask viewers to vote on a topic they care about on your website or other digital space.

Sign-up for our newsletter: Ask viewers to sign up for your newsletter to gain more insight, and to make sure they don't miss announcements.

Share on Social Media: Don't take for granted reiterating to your audience to share on social media to their audience or groups. While this seems commonplace, it is still worth mentioning.

Support your Sponsor: If you have a sponsor as part of your broadcast let your viewers know how to connect and follow that person.

A good call to action is to give a clear and direct command of the next steps. You always want to direct your audience to next steps. This should be concise and convincing. Experts recommend keeping it short in length, such as a short phrase, not a sentence. Surprisingly, most are no longer than five words. Too often I see live video with a solid introduction, great content, but the close falls short, it's weak at best. This means you must understand your message and your conversion goal. Every call to action needs to answer two questions: What do you want viewers to do, and why should they do it? It should have a sense of urgency especially if you are highlighting a product. Test out what words or phrases work with your audience which ones grab the most attention. But whatever you do, make sure you are ending strong with a call to action.

Call to Action= Call to Money:

Why do you want to perfect or practice this call to action technique in your live video? Simply as you become more proficient as a live video CEO, you will be invited to showcase your expertise or brand in other influencer's live video. And when they come calling, they will want to make sure you are ready to connect, engage and make an offer to their audience that does not come off offense or lackluster. Live video has offered me the opportunity to take part in various telesummits, group coaching in other groups, Podcasts and blog interviews. It is a great marketing opportunity. Here are a few reasons: the person who invites you as a guest is giving you exposure to their audience, solidifying you as an expert in your field, and giving you an endorsement of sorts for your product.

Jamilah Corbitt is a beast at live video. I first met her during the hay day of Periscope. She created this movement called 'scopenaked' where she encouraged all of us to be transparent, real on live video. The outcome was amazing results of people taking off the mask and being themselves. She and I reconnected because I intentionally took the same tools you are learning here to connect with her. Not because I wanted anything, but to grow my online community of live video bosses. I connected with her during her live videos, and engaged with her online and offline. Then Jamilah invited me to be a part of her live show because she saw the synergy between both of our messages. The experience was electric. We shared some major nuggets with the audience. At the end, Jamilah asked how people can get in contact with me. This is where the offer comes in. Now let me say this. Remember you are still in someone else's "house." Make sure there is a clear understanding between you and the host. Ask before the interview if it is acceptable to pitch a product, group or service. If you are given permission, you still have to be aware on how you position yourself as not to come off as you are trying to take away from the host. As I tell my children; sharing is caring.

As a result of Jamilah sharing her platform and allowing me to make an offer, a lady reached out to me as she wanted to make more impact on live video. After a free 15 minute discovery call, she decided she wanted to attend one of my VIP strategy sessions. AMAZING!! There are so many other stories similar to this one, as I became to use Live video as a serious marketing tool. We are on live video not just because of any other thing, but we are on live video on purpose for a purpose.

Call to Commitment:

I believe relationships should be reciprocal. There should be an investment on both sides. Your call to action is your audience chance to commit to you. If you deliver powerful content, there is nothing wrong with requiring your audience to take the next steps of working with you or downloading a free PDF. The problem most face is reframing their mind into thinking your live video is a money machine. I never played the slots in Las Vegas, but from what I hear, it is a game one can become addicted to. The casino's bank (no pun intended) on that. They want you never to leave your seat. When I do live video, I desire the same results. I want people to think about me but not in a creepy way. I want my audience to feel connected to me, to engage with me like we are familiar friends and naturally tell themselves I want to work with her. And so far, it has worked. I want you to stop acting like the jilted lover. I'll take whatever my audience gives me. No. I need you to steer the ship, Chief Executive Officer. I need you to commit to yourself, and as you do that your audience will commit to you. Do not fear; there are people out there who want to work with you. Commit to the live video process and commit to consistently making your audience an offer which solves their problem and makes their life better. This commitment is the whole reason behind this book you are reading right now. After being challenged by my publisher to write a book centered on my expertise, I had to commit to the process. The process of commitment even with the responsibilities of being a wife, a mother of three, pregnant with the fourth blessing, a media personality, running a business, ministry, community service. Whew! I think that's it. You get the point. Among all the hats I wear, I had to be committed to myself. I had to practice what I am preaching to you today. When you commit to the value of what you bring, you will no longer allow others not to do the same. As my friend and financial educator Steven Hughes of Know Money Inc. would say: Action Cures All!

Lights, Camera and Action:

Now that you know how to connect, engage, and make an offer to your audience go and test it out, if you have not already. The best way to get the most out of this experience is to go ahead, do it and create your CEO story. Live video is not going anywhere. In fact, as relationship marketing increases, so will the need for live video. This is not a trend folks. This is soon becoming the main way to communicate. Facts tell, but stories sell. And the best way to sell your story is to be in front of the screen, and not behind it. This should now be your new normal. Your story deserves attention.

Here's how we can connect and engage:

You can schedule your own discovery session where you can discuss live video, media, content creation or just chit chat about the book. You can find that at www.therosholive.com/contact. I love meeting new friends, and would love to hear from you.

You can also join my **FREE** Facebook Community where I am dropping some serious gems from live video, traditional media to get before anyone else courses. Head over to Facebook, and join The Rosho Live Media Mastermind community. Make sure to introduce yourself and tell us your story.

And finally, if got a few aha's from this book, please do me a favor. Leave a review on the website you downloaded so others will know, and experience what you did. Sharing is caring, so please share.

Take the Pledge:

After reading this book it may have sparked this immediacy to Go Live. While, I am excited about that, the sobering truth is there is also responsibility which comes with using this platform of live video. We have all heard the horror stories of people texting and driving. We live in a culture of distracted drivers. Sadly, the number of accidents caused by distracted drivers is on the rise. I am asking, **"DO NOT GO LIVE while driving."** Your message is important, but not as important as you arriving to your destination ALIVE!

Will you take the pledge and promise me you'll arrive ALIVE before you GO LIVE!

If you have decided to take the pledge to wait until you have safely arrived at your destination I want to know about it. Let's connect at therosholive@gmail.com.

Made in the USA
Las Vegas, NV
24 June 2021

25345687R00033